Table of Contents

TV Coverage
of the Oil Crises:
How Well Was the Public Served?

Volume II
A Quantitative Analysis
1973-74/1978-79

Edited by
Leonard J. Theberge

(**the media institute**)

Preface

Background: The first oil crisis began in earnest during the 1973 Arab-Israeli War when Arab OPEC members placed an oil embargo on the United States (October 17, 1973), reduced production, and, in the following months, quadrupled prices. The embargo was lifted March 19, 1974, though TV coverage of the effects continued into May 1974. Shortages due to the Iranian Revolution were felt in the United States in November 1978; by December 26, 1978, Iran, the world's fourth largest oil producer, ceased oil production due to its internal revolution. Saudi Arabia exacerbated the shortage by cutting production by two million barrels a day in January 1979. In March 1979 OPEC raised the price of oil by 40%; the price was doubled by year's end. In July 1979 Saudi Arabia upped its production by one million barrels a day, and the shortage and its TV coverage wound down in August 1979.

*   *   *

The 1973–74 and 1978–79 oil crises sparked one of the hottest public policy debates and generated one of the biggest news stories of the decade. Because of this issue's prominence on the national agenda and, therefore, in the news, The Media Institute analyzed the coverage given the oil crises on America's most pervasive and trusted news medium—television network evening news. In the interest of thoroughness, researchers analyzed every pertinent story* broadcast during the periods October 1973—May 1974 (an eight-month period) and November 1978—August 1979 (a ten-month period). Over 39 hours of news—more than 1,400 stories—were examined. Using a quantitative research technique called content analysis, the Institute collected and analyzed, using computers, a massive amount of data which yields a number of significant insights into network news coverage of the oil crises.

In view of the extensive coverage the oil crises received, however, this wealth of data has a broader significance, for it reveals how each network reported on a story of critical importance and great complexity. Moreover, a comparison of the networks from one crisis to the next indicates the changes that have occurred in the news operation of each network, and in television news overall.

This analysis is the second in a three-volume study The Media Institute is publishing on network television evening news coverage of the oil crises. The first study analyzes the substantive content of television's coverage—how the causes, effects and solutions to the crises were presented on the news. The third study offers an economic analysis of the accuracy, relevance and balance of television's portrayal of the oil crises.

These three volumes share a common purpose: to contribute to an understanding of how television reports on a major policy issue; and to identify specific areas in which the quality of television news might be improved.

---

*See Appendix for selection criteria

Executive Summary

The two oil crises of the 1970s (1973–74 and 1978–79) were heavily covered on network evening news programs, which devoted over 1,400 stories to the subject. The fundamental similarities in the two crises allow for a fascinating comparison of network evening news of the early 1970s with the same news programs five years later. The changes are striking. In the second crisis:

- The average story was 34% longer. (see page 4)
- Reliance on the government as a source of information increased 13%. (see page 38)
- Use of government sources on camera increased 32%. (see page 25)
- Government policies received less favorable treatment. (see page 54)
- Coverage had a more economic and less political focus. (see page 50)
- The percentage of stories using film or videotape increased 35%. (see page 10)
- The percentage of oil crisis stories beginning in the first ten minutes of the newscast increased 31%. (see page 4)

Such aggregate assessments do not, of course, reveal the individual character of each network's coverage. The networks mirrored one another in some respects:

- Each devoted one-half of its coverage of solutions to conservation and rationing. (see pages xii–xiv)
- Each gave non-market solutions three times as much coverage as market solutions. (see pages xii–xiv)
- Each devoted less than one-fifth of its coverage of causes to the role of the government as a possible cause. (see pages xvi–xviii)

In other respects, each network had a unique character:

ABC

In several measures, ABC was the pacesetter in the first crisis (1973–74). By the second crisis (1978–79), however, CBS and NBC had surpassed ABC in many respects. ABC:

- Had the lowest percentage of anchorman stories in the first crisis, and the highest percentage in the second. (see page 15)
- Had the highest percentage of stories with film or videotape in the first crisis and the lowest in the second. (see page 12)
- Had the longest average story length in the first crisis. (see page 5)
- Devoted the greatest amount of coverage to on-camera sources in the first crisis and the lowest in the second. (see page 21)
- Referred to government policies most often in the first crisis and least often in the second. (see page 53)
- Gave the man-in-the-street the most coverage in the first crisis and least in the second. (see page 35)
- Used number overlays the most in both crises. (see page 24)
- Gave the most coverage to the regulation vs. deregulation and price controls vs. decontrol debate. (see page xii)
- Showed the strongest preference for covering non-market rather than market solutions. (see page xii)
- Gave most attention to the government as a possible cause of the crises. (see page xvi)

CBS

The role of the anchorman was dramatically reduced by CBS from the first to the second crisis. The second crisis was covered 53% more by CBS than by either of the other two networks. CBS:

- Used anchormen, or "talking heads", for 53% of stories in the first crisis—but this number dropped to 26% in the second crisis. (see page 15)
- Had the lowest percentage of stories using film or videotape in the first crisis and the highest percentage in the second, with a proportionate increase of nearly 60%. (see page 12)
- Gave on-camera news sources the shortest appearances in both crises. (see page 23)
- Had the highest percentage of stories using graphs in the second crisis. (see page 24)
- Had coverage with the most economic and the least political focus. (see page 52)
- Portrayed government policies least favorably. (see page 55)
- Was the most inclined to criticize the government for not enough involvement in the oil crises. (see page 55)
- Showed the strongest preference for covering the oil industry as a possible cause—40% more than NBC. (see page xvii)

NBC

Not only did NBC cover the first crisis most extensively, but it also gave the first crisis intensive coverage at least a month before the other two networks. NBC:

- Devoted 12% more coverage to the first crisis than did either ABC or CBS. (see page 1)
- Had the lowest percentage of stories beginning in the first ten minutes of the newscast in the first crisis, and the highest percentage in the second. (see page 6)

- Relied most on the government as a source of information in both crises. (see page 41)
- Gave foreign news the most coverage in the first crisis. (see page 51)
- Had coverage with the most political and the least economic focus. (see page 52)
- Referred to government policies most often in the second crisis. (see page 53)
- Gave the least coverage to regulations and price controls as a solution to the crises. (see page xiv)
- Showed the strongest preference for covering OPEC as a possible cause—47% more than CBS. (see page xviii)

INTRODUCTION

Volume I of this study analyzed the content of the news coverage given the two oil crises of the past decade (1973–74 and 1978–79) on the television networks' evening news broadcasts. No distinction was made among the three networks in that volume regarding the quality of their coverage. Did they cover the oil crises in uniform fashion? That is, was there no notable difference among them, or was the viewer best served by ABC, CBS or NBC?

A look at the oil crises solutions that received coverage suggests the viewer heard pretty much the same options on each network. Volume I reported that 50% of the discussion of solutions focused on conservation and rationing. As Charts A, B and C show, the figure was 51% for ABC, 49% for CBS, and 50% for NBC. Volume I also revealed that non-market solutions (regulation, price controls, conservation, rationing, divestiture of oil companies and countermeasures against OPEC) received three times the coverage given to market solutions (decontrol, deregulation, reliance on price mechanisms and domestic oil production). This three-to-one preference for non-market solutions was reflected in both the CBS and NBC coverage. ABC's preference was nearly four to one. The debate on regulation vs. deregulation received slightly more attention on ABC (17%) and CBS (16%) than on NBC (13%).

From this data, one might conclude that the three net-

Chart A

Solutions Discussed on ABC

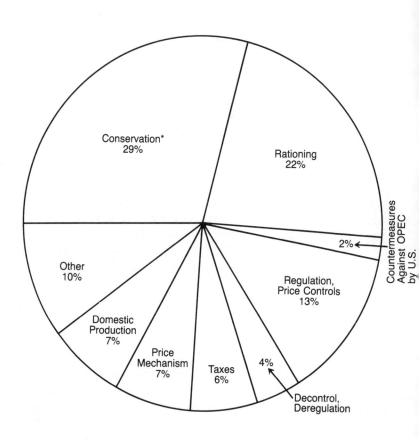

*excluding price-induced conservation

Chart B

Solutions Discussed on CBS

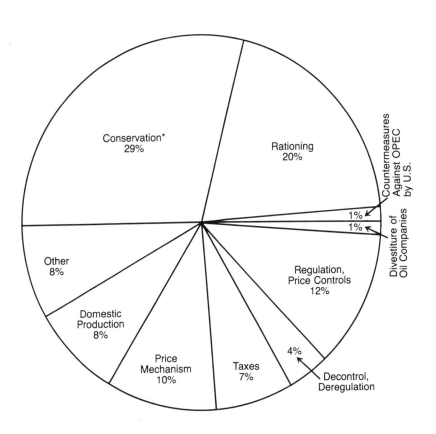

*excluding price-induced conservation

Chart C

Solutions Discussed on NBC

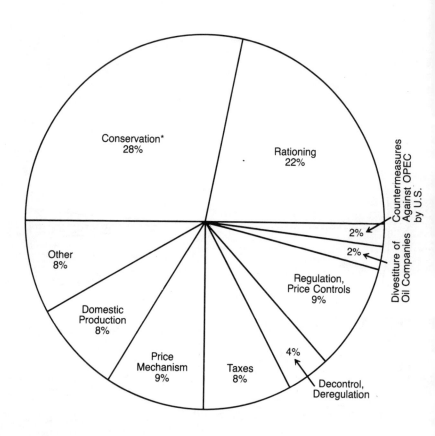

*excluding price-induced conservation

works gave the oil crises very similar treatment. Further study, however, indicates that this was by no means the case. Charts D, E and F show that each network covered the causes of the oil crises in distinctly different fashions. In particular, they diverged markedly in the relative importance given to OPEC and the oil industry as possible causes. NBC devoted much more coverage to OPEC as a possible cause (44% of NBC's coverage of causes) than to the oil industry (30%). Quite the reverse was true at CBS, where the oil industry's role as a cause was given much more attention. Of all the coverage CBS gave to the causes of the crises, the oil industry was the focus of 42% of the coverage, and OPEC just 30%. On ABC, the two possible causes received nearly equal amounts of coverage (OPEC, 35%; oil industry, 34%).

On the other hand, the three networks were in close agreement concerning the importance of the government as a possible cause. Both CBS and NBC devoted 18% of their coverage to government as a cause, while ABC gave 19%.

In this volume, the differences and similarities among the three networks will be explored in great depth. Which news program, for example, is most likely to carry information originating from Congress? Which broadcasts the most man-in-the-street interviews? Which network is most apt to show a graph in a story? The differences in some cases are striking.

This volume compares the networks' news programs in a variety of ways. The first chapter presents an overview of the amount of coverage given each oil crisis by each network, and examines some of the more mechanical aspects of the news programs, such as the placement of stories in the newscast. The second chapter delves further into such production questions, including the relative use of spokesmen, videotape, stand-uppers, graphs, etc.

The third chapter addresses questions which are more journalistic in nature. Who or what is the source of the news? Who, other than reporters, is actually shown speaking on TV (termed here spokesmen)? What percentage of spokesmen come from government? Are spokesmen from Congress or from the oil industry featured more promi-

Chart D

Causes of the Oil Crises Discussed on ABC

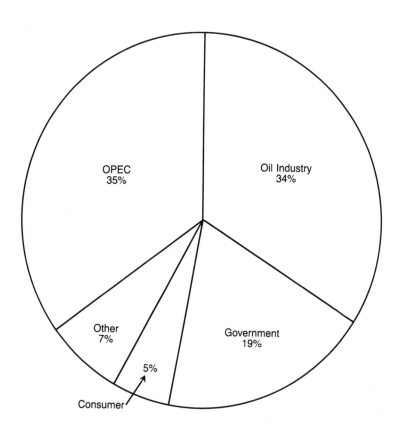

OPEC
35%

Oil Industry
34%

Other
7%

5%

Consumer

Government
19%

Chart E

Causes of the Oil Crises Discussed on CBS

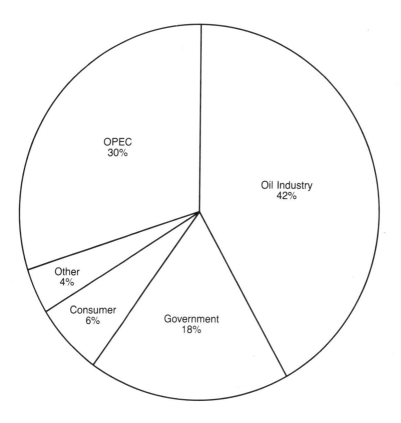

OPEC
30%

Oil Industry
42%

Other
4%

Consumer
6%

Government
18%

Chart F

Causes of the Oil Crises Discussed on NBC

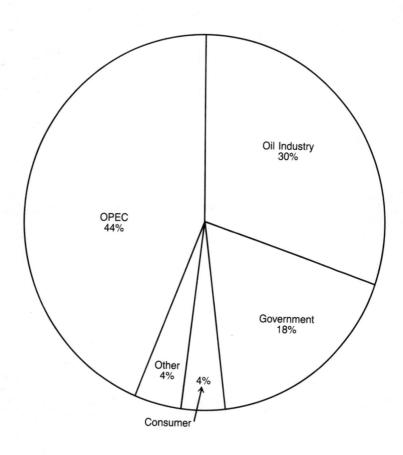

nently on the news? Finally, the fourth chapter examines, by network, some substantive content questions like those addressed in Volume I.

One note of caution: because the crises were not identical, changes in news coverage from one crisis to the next may reflect this difference in events as well as differences in the networks' reporting techniques.

Chapter I

The three major television networks' evening news programs devoted a great deal of attention to the two oil crises of the 1970s. For this study, a total of 39½ hours of videotape was analyzed. Of this total, 61%, or over 24 hours, occurred in the eight months of the first oil crisis. The second oil crisis received more than 15 hours of coverage over a period of ten months.

The three networks varied in some surprising ways in the amount of air time each gave to the crises. In the first crisis, NBC devoted the most time to the story, with 12% more coverage than the others—almost one hour more. In the second crisis, CBS allotted far and away the most time to covering the oil crisis, with 53% more coverage than either of the other two networks. ABC and NBC were remarkably similar in the amount of total coverage each gave the second crisis, varying in the aggregate by less than a minute. [See Graph 1]

Another way to measure coverage is in terms of the number of stories broadcast. [See Graph 2] In total, 1,462 oil stories were broadcast by the three networks during the two crises. Measuring coverage in this fashion yields an overall picture similar to that based on coverage measured in hours. The differences are interesting, however, for they indicate that the average story length varied from one crisis to the next, and among networks. In the first crisis, the average

Graph 1

Coverage of Oil Crises
Measured in Hours

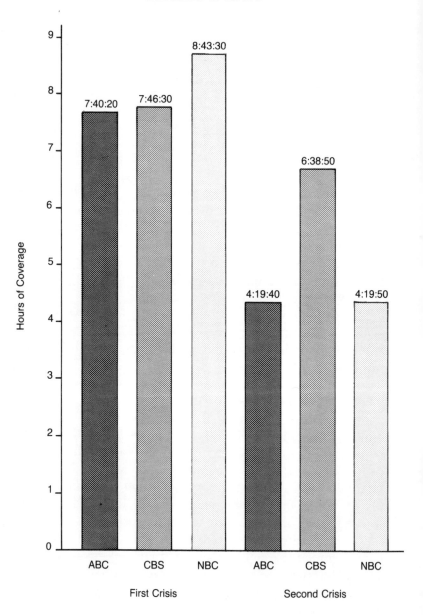

Graph 2

Coverage of Oil Crises
Measured in Stories

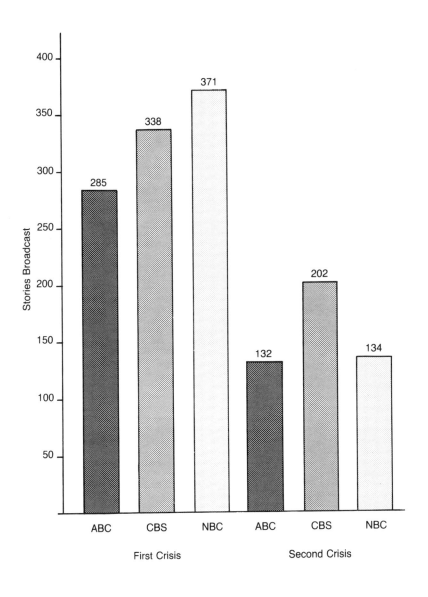

duration of an oil crisis story was 88 seconds. In the second crisis, the average story was 34% longer, at 118 seconds. Among networks, the variance was most pronounced during the first crisis, in which the average ABC story was 14% longer than the average story of either CBS or NBC. [See Graph 3]

A third way to measure the coverage of the oil crises is in terms of the placement of stories within a broadcast. The half-hour network evening news broadcasts generally begin with what is deemed to be the most important news of the day, and frequently conclude with more light-hearted fare.* Over 60% of all oil crisis stories began in the first ten minutes of the program, and one in five of these was the lead story. Another 32% started in the middle ten minutes, while only 7% of all oil crisis stories began in the final ten minutes.

In the first crisis, NBC was the only network with less than 50% of oil stories beginning in the first ten minutes of the broadcast. On the other hand, it had the highest percentage of lead stories, while ABC had the lowest. Coverage of the second crisis was even more heavily skewed toward the beginning of the broadcast, with a 31% increase in the proportion of stories beginning in the first ten minutes. NBC had both the highest percentage of lead stories (20%) and the highest percentage of stories in the first ten minutes (82%). ABC had the lowest percentage of second crisis stories beginning in the first ten minutes, as well as the highest percentage of stories in the final ten-minute segment. [See Graph 4]

Coverage of the first crisis differed from that of the second not only in absolute quantity, but also in intensity. The two-month period of December 1973—January 1974 accounted for over eleven hours of coverage or 47% of the total coverage of the first crisis. Over half of the air time on some news broadcasts during this period dealt with the oil crisis alone, although on average oil stories accounted for about 22% of news time during this two-month period. The

---

*It should be noted, however, that the audience of the evening news increases over the course of the broadcast, so that the final stories have the largest audience.

Graph 3

Average Length of Oil Crisis Stories

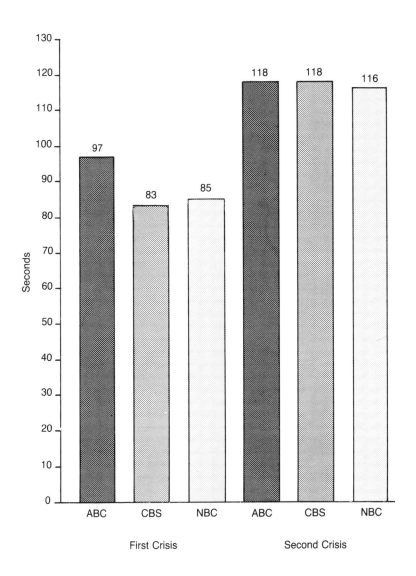

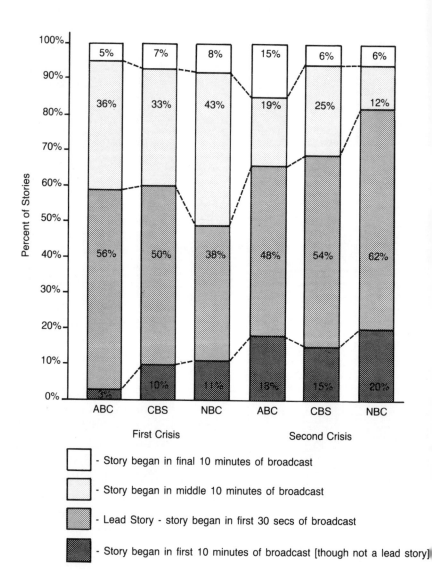

Graph 4

Placement of Oil Crises Stories
*[100% = total number of
stories broadcast by a network
during one crisis]*

Percent of Stories

	ABC	CBS	NBC	ABC	CBS	NBC
	5%	7%	8%	15%	6%	6%
	36%	33%	43%	19%	25%	12%
	56%	50%	38%	48%	54%	62%
	3%	10%	11%	18%	15%	20%

First Crisis   Second Crisis

☐ - Story began in final 10 minutes of broadcast

▨ - Story began in middle 10 minutes of broadcast

▨ - Lead Story - story began in first 30 secs of broadcast

■ - Story began in first 10 minutes of broadcast [though not a lead story]

Graph 5

Coverage of Oil Crises Per Month:
Combined Coverage of Three Networks

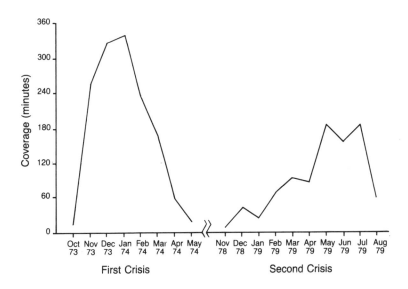

second crisis was not as clearly demarcated; coverage increased at a much slower rate in the beginning months of the second crisis than it did in the first. There were two separate months of peak coverage, May and July 1979, with a slackening of coverage during June 1979. [See Graph 5]

In addition, during each crisis each network's coverage peaked in a different month. In the first crisis, NBC was the first to cover the story intensively, with 66% more coverage than either CBS or ABC in November 1973. NBC's coverage decreased somewhat in December 1973, while CBS' coverage peaked that month. ABC's coverage did not peak until January 1974, well after the other two networks. From January through April 1974, however, coverage decreased at a virtually uniform rate among all three networks.

Such uniformity is in sharp contrast to the widely di-

Graph 6

Network Coverage of Oil Crises
Per Month

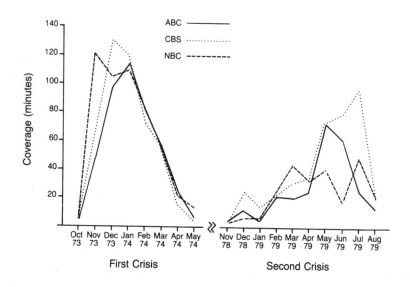

vergent coverage of the second crisis. NBC's coverage was intensive in May and July 1979, with a slack interim month. ABC, on the other hand, gave its most intensive coverage to the second oil crisis in May 1979, and its coverage declined sharply after that. In stark contrast, CBS' coverage of the second crisis reached its peak in July 1979, at a level of coverage double that of NBC and more than triple that of ABC. [See Graph 6]

Chapter II

The evening news programs of the television networks are the product of large and highly skilled news organizations working under considerable pressure. The actual broadcast is the culmination of innumerable decisions made by producers, writers, correspondents, technicians and others. Beyond the question of the quantity of coverage each story merits, decisions concerning the quality of coverage must also be made. These questions include the amount of time and money to be spent developing a particular story; the best mix of studio stories ("talking heads") and stories with footage (film or videotape); the optimum length for a story that has no footage, and so forth.

Such questions confront network news organizations on a continuous basis, and each network has a different set of answers. It appears, moreover, that the answers have evolved over the course of the decade, due in no small part to technological innovations. Thus a comparison of the different ways the networks have approached these questions suggests something about the direction of television news in general, in addition to showing differences among the networks.

One important qualitative measure of news coverage concerns the production techniques employed. It is obviously less difficult to have someone read the news while seated before a camera than it is to send a camera to the site of a news story, relay that footage back to the studio, edit and broadcast it, all before the news has become "stale". However, such up-to-the-minute footage is the unique feature distinguishing television from all other news media, and is its biggest draw.

Of all oil crisis stories broadcast, over half had footage of some sort, either film or videotape. The percentage of stories with footage increased markedly from the first crisis to the second, by over 35%.* [See Graph 7] In the first crisis, ABC had a much higher percentage of film stories than did either CBS or NBC. By the second crisis, the situation was exactly reversed. ABC, despite an increase in the proportion of its stories with footage, had sunk to last place. CBS, on the other hand, went from last to first place, increasing by almost 60% the proportion of film stories it broadcast. [See Graph 8]

The increase in film stories from the first crisis to the second corresponds with the decline in anchor stories, the so-called "talking head." The proportion of stories with only an anchorman declined by 40% from the first crisis to the second. [See Graph 9] In the first crisis, CBS had the highest percentage of anchor stories—53%. This was cut in half by the second crisis, when anchor stories accounted for only 26% of all CBS stories. As might be expected from the data concerning film stories, ABC changed in this regard the least, yet went from having the lowest proportion of anchor stories in the first crisis to having the highest proportion in the second. [See Graph 10]

A third type of story, the stand-upper, is a sort of hybrid of the other two in that it is not filmed in the studio, yet it shows little more than a reporter and whatever he is standing in front of—the White House or some other backdrop. While stand-uppers are frequently the opening or closing for a film segment, they are rarely used by themselves, and account for only 2% of all stories. Use of stand-uppers increased in the second crisis for all three networks, account-

---

*N.B.—When percentages are compared in this study, their difference is expressed as the relative percentage difference, rather than the absolute percentage difference. In this example, the percentage of stories with film or videotape increased from 51% to 69%, a relative increase of 35%, though an absolute increase of 18%.

Graph 7

Percent of Oil Crisis Stories
With Film or Videotape
[100% = total number of stories broadcast by all three networks during one crisis]

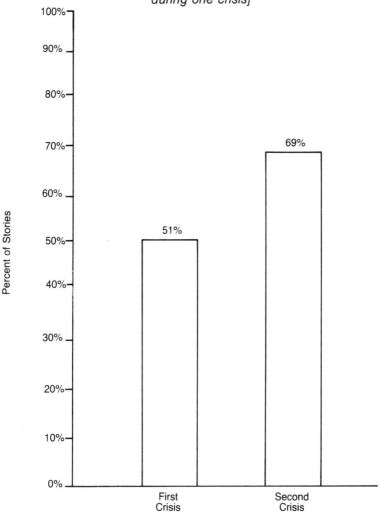

Graph 8

**Percentage of Oil Crisis
Stories With Film or Videotape**
*[100% = total number of stories
broadcast by a network
during one crisis]*

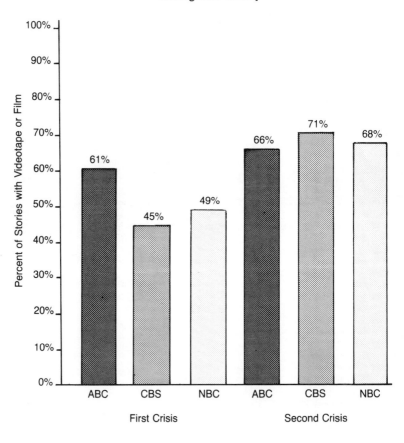

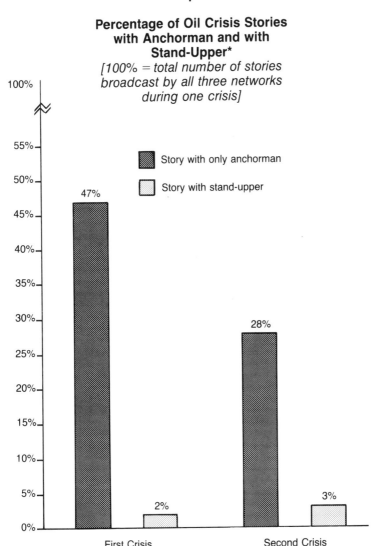

Graph 9

**Percentage of Oil Crisis Stories
with Anchorman and with
Stand-Upper***

*[100% = total number of stories
broadcast by all three networks
during one crisis]*

Story with only anchorman

Story with stand-upper

47%

28%

2%

3%

First Crisis

Second Crisis

*A "stand-upper" is a story that includes footage of a reporter in the field,
but that has only the reporter (and background) shown. Such stories
frequently include anchorman lead-ins.

ing for 5% of NBC's second oil crisis stories. [See Graph 10]

All stories, however, are not created equal. Anchor stories are, on the average, under thirty seconds in length, while film stories average around two and one-half minutes. Stand-uppers average somewhere in between, generally just over a minute and a quarter in length.

Anchor stories averaged a few seconds less in the second crisis than in the first, with the 16% decrease in length on CBS accounting for most of the decline. Film stories, on the other hand, increased in average length in the second crisis. CBS had the briefest film stories in both periods, but the differences were not great. In the stand-upper category, ABC showed a penchant for rather brief stories in the first crisis, and for extremely long ones in the second. In both periods, however, stand-uppers were only a small percentage of the total. [See Graphs 10 and 11]

Another way to consider the relative importance of anchor, stand-upper and film stories is in terms of the total time each type of story was broadcast. Although 41% of all stories were anchor stories, the relative brevity of this type of story meant that only 12% of total coverage, as measured in minutes, was in the form of anchor stories. Because the relative proportion of anchor stories declined from the first to the second crisis, while the duration of the average anchor story also declined, it is not surprising that anchor stories accounted for a lower percentage of the total coverage in the second crisis. The size of the decline—over 50%—is striking, however. Such a comparison also highlights how heavily CBS relied on its anchorman (usually, though not always, Walter Cronkite) to report the news in the first crisis. Twenty percent of all CBS's coverage of the first oil crisis was presented by the anchorman, as compared to 15% for NBC, and 10% for ABC. In contrast, during the second crisis the three networks presented almost identical proportions of their coverage in anchor stories (about 6%). Stand-uppers accounted for a slightly increased percentage of coverage in the second crisis, but the role of such stories was minimal throughout. [See Graph 12]

Film stories, therefore, constitute the lion's share of air time. In the first crisis, 83% of all coverage took the form of

Graph 10

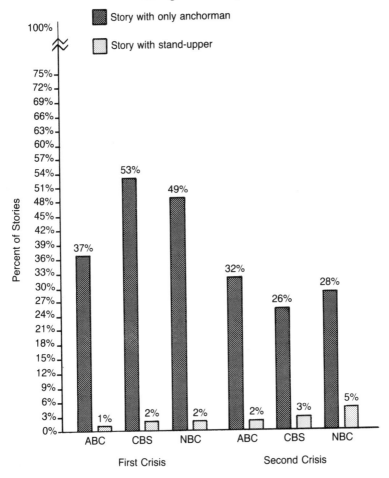

**Percentage of Oil Crisis Stories
with Anchorman and with
Stand-Upper***
*[100% = total number of stories
broadcast by a network
during one crisis]*

■ Story with only anchorman

□ Story with stand-upper

Percent of Stories

| 100% |
| 75% |
| 72% |
| 69% |
| 66% |
| 63% |
| 60% |
| 57% |
| 54% |
| 51% |
| 48% |
| 45% |
| 42% |
| 39% |
| 36% |
| 33% |
| 30% |
| 27% |
| 24% |
| 21% |
| 18% |
| 15% |
| 12% |
| 9% |
| 6% |
| 3% |
| 0% |

First Crisis: ABC 37%, 1% — CBS 53%, 2% — NBC 49%, 2%

Second Crisis: ABC 32%, 2% — CBS 26%, 3% — NBC 28%, 5%

First Crisis   Second Crisis

*A "stand-upper" is a story that includes footage of a reporter in the field,
but that has only the reporter (and background) shown. Such stories
frequently include anchorman lead-ins.

Graph 11

Average Length of Oil Crisis Stories

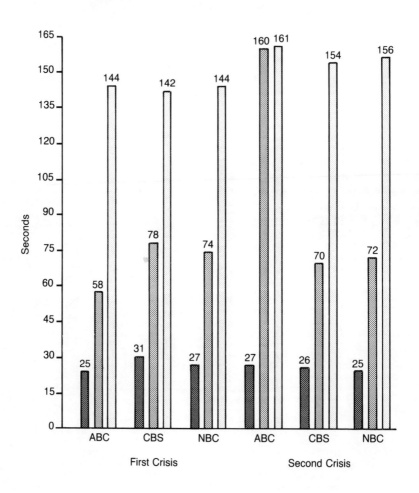

■ Anchorman stories

▨ Stand-upper stories

□ Stories with film or videotape

Graph 12

Percent of Total Coverage Accounted for by Anchorman Stories and by Stand-Uppers
[100% = total coverage, measured in minutes, by a network during a crisis]

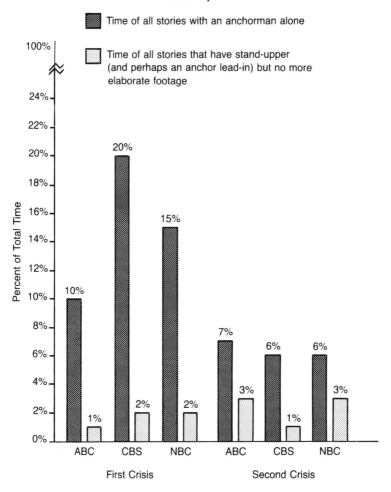

Time of all stories with an anchorman alone

Time of all stories that have stand-upper (and perhaps an anchor lead-in) but no more elaborate footage

Percent of Total Time

	First Crisis			Second Crisis		
	ABC	CBS	NBC	ABC	CBS	NBC
Anchorman alone	10%	20%	15%	7%	6%	6%
Stand-upper	1%	2%	2%	3%	1%	3%

stories with film. Here again, ABC had the highest proportion, and CBS the lowest. In the second crisis the proportion was still higher, at an average of 91% and, as in the other related measures, CBS had taken the lead while ABC held last place. [See Graph 13] However, this measure does not indicate the total amount of film contained in a story, for a film story often included a lead-in from the anchor, possible stand-uppers from the correspondent, and perhaps a closing from the anchor as well.

A different measure of the use of film concerns spokesman appearances. For the purposes of this study, a spokesman is any non-newsman who is both seen and heard saying something in a story. Nearly four hours worth of spokesmen appeared in coverage of the first crisis, and another two hours appeared in the second. ABC used spokesmen most extensively in the first crisis, while CBS had by far the most spokesmen in the second. [See Graph 14] ABC's extensive use of spokesmen in the first crisis is particularly interesting in that ABC's total coverage of the first crisis was less than that of either CBS or NBC. [See Graph 1] As a percentage of total coverage, spokesman appearances accounted for 18% of all ABC's first crisis coverage. In contrast, NBC, which covered the first crisis the most thoroughly of the three networks, had the lowest reliance on spokesmen, at 14% of total coverage. [See Graph 15] Although the relative importance of film stories increased from the first to the second crisis [See Graphs 7 and 13], the percent of coverage in the form of spokesmen remained constant for both CBS and NBC. ABC, on the other hand, reduced dramatically the use of spokesmen appearances, by one-third, so that spokesmen received the least coverage in the second crisis on ABC. [See Graph 15]

The average spokesman was given a significantly shorter appearance in the second crisis than in the first. While in the first crisis the average spokesman appearance had a duration of 20 seconds, he was given only a 15-second bite in the second crisis. NBC had the longest spokesman appearances in the first crisis, averaging 22.6 seconds, and CBS the shortest, with an average of 18.5 seconds. In the second crisis, all three networks were in surprisingly close agree-

Graph 13

Percent of Total Coverage Accounted For
Stories that Included Film or Videotape*
*[100% = total coverage, measured in
minutes, by a network during
one crisis]*

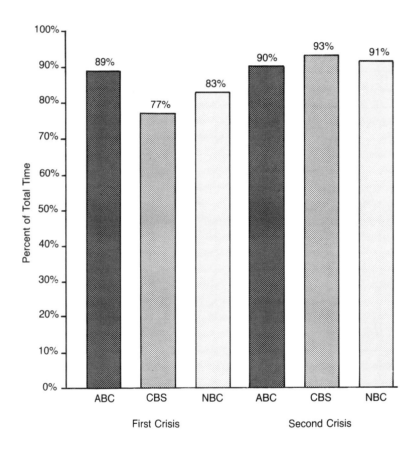

*The total length of such stories is included here, though they usually have
an anchorman lead-in, and often a stand-upper portion.

Graph 14

Total Minutes Spokesmen*
Appeared in Oil Crisis Coverage

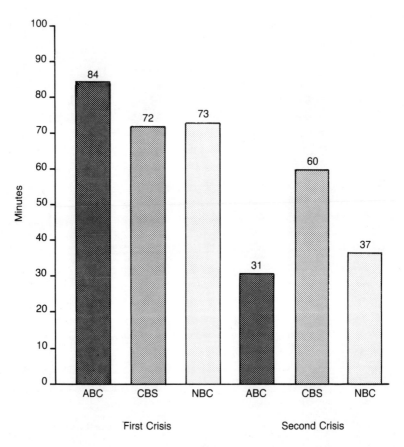

*"Spokesman" refers to any non-newsman both seen and heard saying
something in a story.

Graph 15

**Percentage of Total Coverage
Accounted for by Spokesman Appearances**
*[100% = total minutes of coverage
broadcast by a network
during one crisis]*

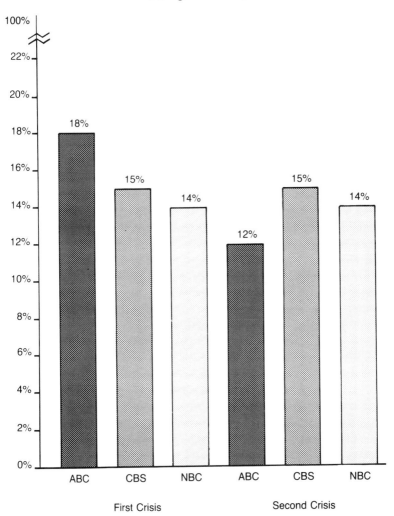

ment, though CBS continued to have the shortest appearances. [See Graph 16]

One other way to measure the production efforts of the networks is in their use of graphics to portray statistical data. The oil crises were difficult stories to report because they involved many issues of business and economics where the real story might lie in a statistic, or a statistical trend. Though some statistics speak for themselves, they never appeared as spokesmen. Rather, such data were at times flashed on the screen as a number overlay, or at other times presented in the form of a graph. Overall, 89% of the oil crisis stories had no such graphics. Less than 2% had graphs, and about 9% had number overlays. Both types of graphics appeared in a much higher percentage of stories in the second crisis than in the first. In the first crisis, NBC used graphs the most, though nonetheless in under 1% of all its stories. CBS took the lead in this measure in the second crisis, with over 5% of its stories accompanied by graphs. ABC came in last, at about half CBS' level of use.

Number overlays (i.e. numbers appearing on the screen behind the anchorman) were used much more often than were graphs. In the first crisis, ABC had the highest percentage of stories with number overlays, and NBC the lowest. In the second crisis, ABC increased its proportional use of number overlays by 70%, and remained the network using such graphics the most. NBC more than doubled its use of number overlays, and was second in this use among the three networks in the second crisis. CBS also increased the percentage of its stories that had number overlays, but by a more modest jump, and had the lowest level of use of the three. [See Graph 17]

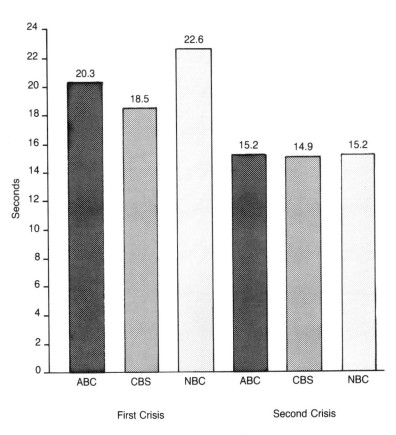

Graph 16

Average Duration of Spokesman Appearances* in Oil Crisis Stories

First Crisis

Second Crisis

*Note: When a spokesman appeared twice in the same story, the two appearances were aggregated.

Graph 17

Visual Presentation of Statistics in Oil Crisis Coverage

*[100% = total number of stories
broadcast by a network
during one crisis]*

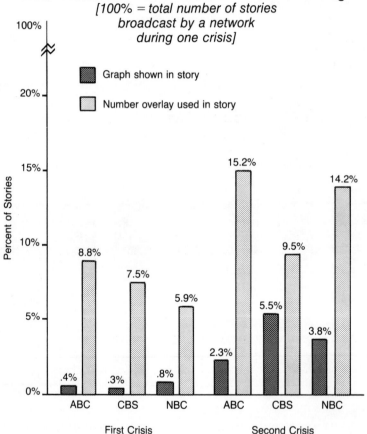

Chapter III

The information reported on the networks' evening news broadcasts is gathered from a wide variety of sources. The sources for a particular story are chosen on the basis of such variables as the nature of the story, the availability of information and the reporter's particular set of contacts. In oil crisis stories, on-camera sources (termed here spokesmen) were from the government 56% of the time. In the first crisis, 50% of these spokesmen were government employees, while in the second crisis 66% were, a proportionate increase of 32%. ABC relied least on government spokesmen, but the differences among the networks were not great. In the second crisis, NBC used government spokesmen 72% of the time, 16% more than CBS. [See Graph 18]

The networks' use of spokesmen from the federal government (including the Administration and all federal departments and agencies, but excluding Congress, which will be considered separately below) accounted for 32% of all spokesman appearances. In the first crisis NBC was the most reliant on the federal government and CBS the least. In the second crisis NBC used the federal government proportionately less than the other two networks. CBS maintained a level of use consistent with that of the first crisis, while ABC increased its use of federal government spokesmen, using them more often than did either CBS or NBC.

In part, the federal government accounted for such a

Graph 18

**Percent of All Spokesmen From
Government [Congress, Federal, State & Local]**
*[100% = total time spokesmen
were on one network in
one crisis—see graph 12]*

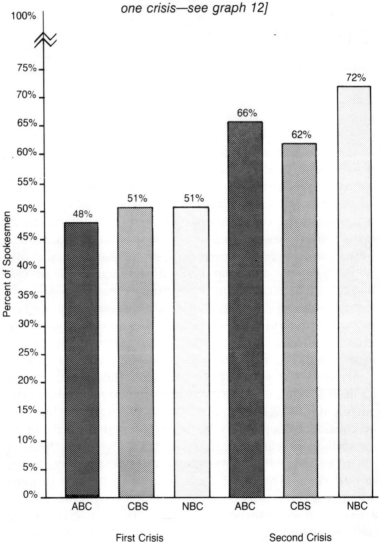

high percentage of total spokesman time because, on the average, such sources were given a longer "bite" than any other category of spokesman. In the first crisis the average spokesman was allotted 20 seconds per story; in that same crisis the average spokesman from the federal government was allotted 27 seconds. In the second crisis, with the overall average down to 15 seconds, the federal government spokesmen averaged 20 seconds. NBC was the most generous with its time in this respect in the first crisis, yet the least so in the second. CBS provided approximately one second more per appearance than did ABC in both crises. [See Graph 19]

While the federal government's total spokesman time declined overall by about 5% from the first crisis to the second, Congress' portion more than doubled. While Congressional spokesmen accounted for 12% of all spokesman coverage in the first crisis, they accounted for 30% in the second. The increase was most dramatic on NBC—an increase of 320%. In the first crisis, NBC gave the least attention to Congress, according it only 9% of all spokesman time. In the second crisis NBC was the most enthusiastic in its coverage of Congressional spokesmen, who accounted for 38% of NBC's total spokesmen. On CBS, Congressional spokesman sources comprised 16% of all spokesman coverage in the first crisis, the highest share of the three networks. This increased 62% in the second crisis, yet CBS gave proprotionately the least coverage to Congress in the second crisis. ABC in both periods was somewhere between the other two networks in this respect.

As with federal government spokesmen, those from Congress were allowed longer appearances on the average than were spokesmen on the whole. In both crises, NBC was the most liberal in this respect and CBS the most parsimonious. [See Graph 20]

The remaining government spokesmen were from state or local governments. The use of such spokesmen increased just slightly in the second crisis, with ABC the least disposed, in both crises, to present such officials. Their appearances were of a duration very close to that of spokesmen in general. [See Graph 21]

Graph 19

**Percent of All Spokesmen From The Federal Government
(Excluding Congress)**
*[100% = total time spokesmen
were on one network
in one crisis]*

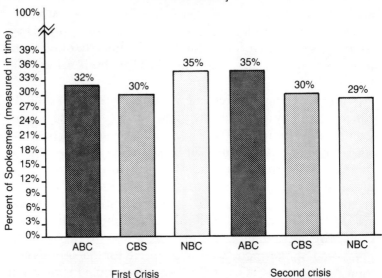

Average Length of Spokesman Appearance:
Federal Government (Excluding Congress)

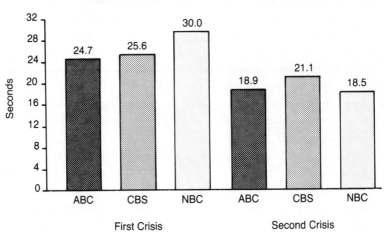

Graph 20

Percent of All Spokesmen From Congress
[100% = total time spokesmen were on one network in one crisis]

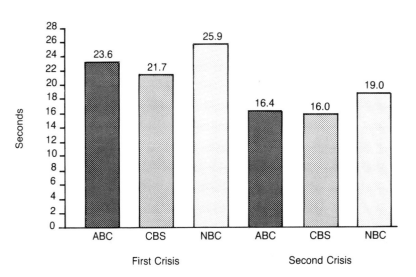

Average Length of Spokesman Appearance: Congress

Graph 21

Percent of All Spokesmen From State & Local Government
*[100% = total time spokesmen
were on one network
in one crisis]*

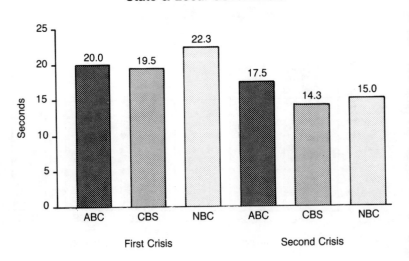

Average Length of Spokesman Appearance:
State & Local Government

Oil industry spokesmen accounted for 14% of all coverage of spokesmen. The coverage was higher in the first crisis, at 15%, and declined to 12% in the second. NBC gave such spokesmen the most coverage in the first crisis, and the least in the second. ABC, on the contrary, went from last to first place in this respect. The substantial amount of NBC's first crisis coverage is explained in part by the long appearances it permitted oil industry spokesmen. NBC gave such spokesmen less than half as much time, on the average, in the second crisis. Overall, the average appearance for oil industry spokesmen was shortened by 43% from the first to the second crisis. Thus, while the average appearance of oil industry spokesmen in the first crisis was approximately the same length as the average for spokesmen overall, in the second crisis it was three and one-half seconds briefer than the overall average. [See Graph 22]

Spokesmen of the Organization of Petroleum Exporting Countries (OPEC) had a low profile in both crises. CBS had the least coverage of OPEC representatives, accounting for only 1% of CBS' total coverage of spokesmen in both crises. ABC devoted 2% of its spokesman coverage to OPEC in both crises, while NBC went from 4% in the first crisis to 2% in the second. The average appearance of OPEC spokesmen was greater in length than the overall length of appearances for spokesmen in both crises. NBC was particularly generous in this type of coverage in the first crisis, as was CBS, in turn, in the second. [See Graph 23]

A device frequently used in television news is presenting one or two individuals as representative of a much larger group. As an example, if autoworkers are laid off, a typical news story will focus on the plight of an individual autoworker. This serves to personalize the news and generally has a greater emotional impact than would abstract statistics such as an unemployment rate. When the opinions or sentiments of the general public are sought, this device of personalizing the news produces the man-in-the-street interview. The person interviewed is not identified as a government or industry employee, for example, but rather as a person in a gas line or one faced with a steep heating oil bill.

As the public's reaction to the oil crises was a part of the

Graph 22

Percent of All Spokesmen From the Oil Industry
*[100% = total time spokesmen
were on one network
in one crisis]*

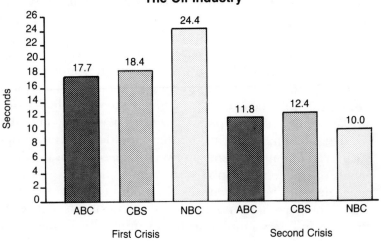

Average Length of Spokesman Appearance:
The Oil Industry

Graph 23

Percent of All Spokesmen From OPEC
*[100% = total time spokesmen
were on one network
in one crisis]*

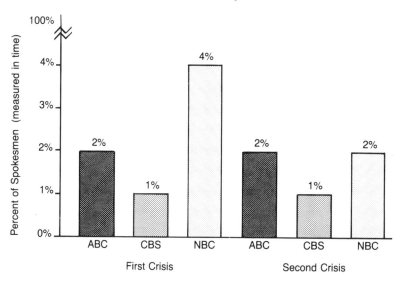

Average Length of Spokesman Appearance:
OPEC

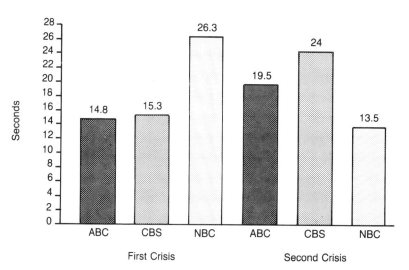

story, man-in-the-street interviews accounted for almost 8% of the total coverage of spokesmen. Such spokesmen were particularly favored in the first crisis, where they accounted for over 9% of all spokesman coverage. This share was halved in the second crisis. ABC used such spokesmen the most in the first crisis, with man-in-the-street spokesmen accounting for 11% of all coverage, compared with 8% for both CBS and NBC. Oddly enough, ABC virtually abandoned this device in the second crisis.

The man-in-the-street was rarely given long to speak. He averaged under 12 seconds in the first crisis, and five seconds in the second. In both crises, he was given an appearance much shorter than the average for spokesmen. [See Graph 24]

One-fifth of all coverage of spokesmen was devoted to spokesmen who fit none of the above categories. They came instead from a wide variety of categories: non-oil industries, labor (particularly truckers and coal miners), experts in economics and finance, consumer and environmental groups, etc. For the purposes of this analysis, they were aggregated as "other" spokesmen. None of these "other" spokesman types dominated the "other" category in terms of coverage allotted. [See Graph 25]

Spokesmen were only one type of news source. The methodology of this study was arranged so that sources not shown on camera were nonetheless tabulated. Whenever a subject (such as airline flight cutbacks or opposition to rationing) was raised, researchers noted the source of the information. The average story could be summed up with about four such subjects. Since each subject is attributed to a source, an equal number of sources and subjects were found in each story. If the source raised several distinct subjects in a story, he was counted several times.

Researchers identified 5,620 subjects, and thus sources of subjects, in the oil crisis stories. On the average a subject summed up 25 seconds of coverage. [See Graph 26, compare with Graph 1] (The content of the oil crisis stories measured in this way is the subject of the first study, and will not be discussed here.)

Graph 24

Percent of All Spokesmen Presented as Man-in-the-Street
[100% = total time spokesmen were on one network in one crisis]

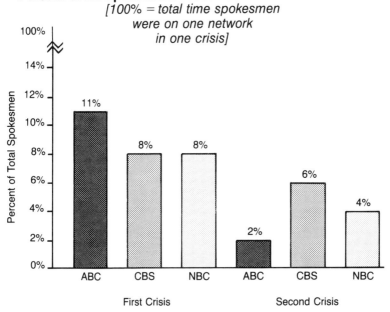

Average Length of Spokesman Appearance: Man-in-the-Street

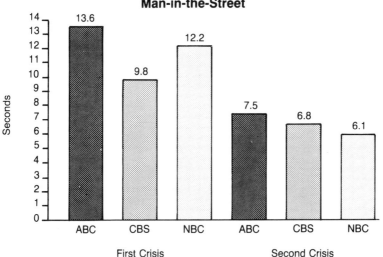

Graph 25

Percentage of All Spokesmen in
Remaining Categories—Here Aggregated as Other*
*[100% = all identifiable sources
of subjects on a network
in one crisis]*

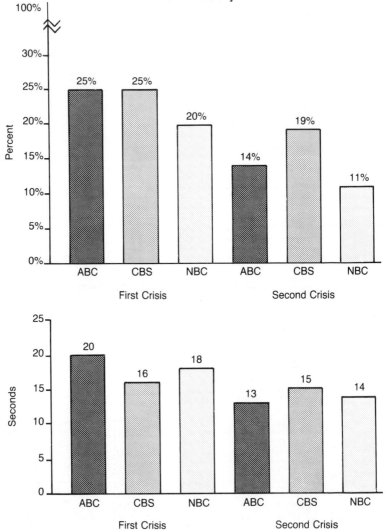

*"Other" includes spokesmen of non-oil industries and businesses, labor, consumer and environmental spokesmen, experts in economics, etc, and those spokesmen not fitting in any category provided the researcher.

Graph 26

Total Number of Subjects
Raised in Oil Crisis Stories

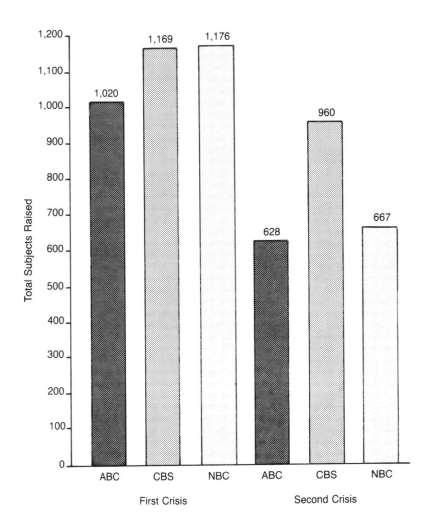

The question of interest here surrounds the type of sources supplying the networks with information. Frequently this was easy to determine. Spokesmen were clearly sources of information and generally fit into one of the categories already discussed. When a source was quoted or identified (even generally as "White House sources said today," for example), the task was simple. At other times, the source was not explicitly identified, yet could be determined with some certainty. Nonetheless, 18% of the sources could not be identified with confidence. At times this was because the reporter was editorializing or stating a commonly held position (*e.g.*, "Whatever happens, the price is certain to rise.") At other times, the story could have come from more than once source. Whenever a subject had an unattributed source, and the source could not be easily inferred by the researcher, the source was coded as unidentifiable. Interestingly, this was more often the case in the second crisis than in the first, with ABC particularly hard to pin down in the second crisis. The remaining sources—82% of the total—could be identified. [See Graph 27]

The prominence of the government as a source was consistent with its standing as a spokesman. Government sources accounted for 56% of all identified sources. The ratio increased from 53% in the first crisis to 60% in the second, a proportionate increase of 13%. In both periods, NBC relied most heavily on the government for information. CBS used government sources the least in the first crisis and ABC used them least in the second. [See Graph 28]

The federal government was used more often as a source in the second crisis than in the first, despite the fact that for the federal government as a spokesman, the trend was the reverse. [See Graph 19] ABC was the most dependent on federal government sources in the first crisis, and the least so in the second. CBS, on the other hand, used federal government sources the least in the first crisis, while NBC was most reliant on such sources in the second crisis. [See Graph 29]

The importance of Congress as a source increased by nearly 50% from the first to the second crisis. CBS was the

Graph 27

Percent of Sources That Were Identifiable (i.e., Percent of All Subjects for Which the Source Could Be Identified)
[100% = all the subjects* raised on a network in one crisis]

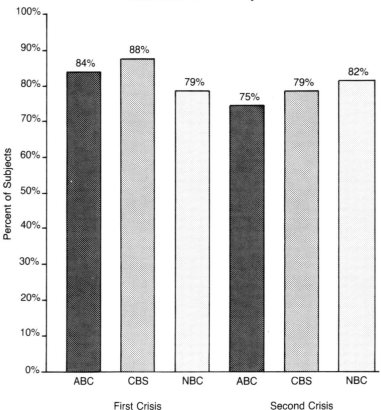

*N.B. every subject had a source, and vice-versa.

Graph 28

Percentage of All Identifiable Sources From the Government
*[100% = all the identifiable sources
of subjects on a network
in one crisis]*

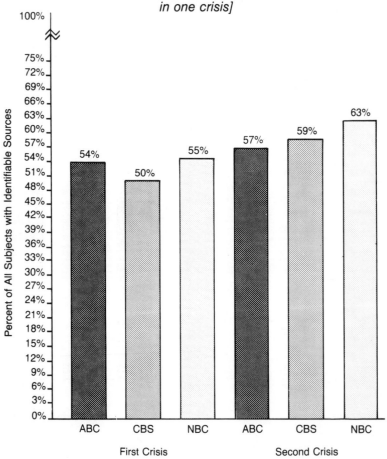

Percent of All Subjects with Identifiable Sources

	First Crisis			Second Crisis	
ABC	CBS	NBC	ABC	CBS	NBC
54%	50%	55%	57%	59%	63%

Graph 29

Percentage of Identifiable Sources From the Federal Government (Excluding Congress)
[100% = all the identifiable sources of subjects on a network in one crisis]

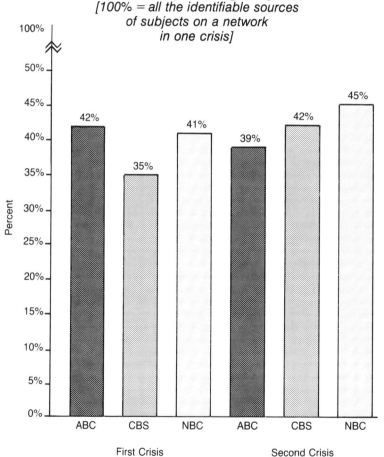

First Crisis   Second Crisis

most reliant on Congress in the first crisis, yet the least so in the second. ABC used Congress the least in the first period, while NBC used Congressional sources the most in the second. Overall, however, the networks did not vary greatly among themselves in either crisis in this respect. [See Graph 30]

Use of sources from state and local governments was relatively constant, at about 4% of all identifiable sources. ABC and NBC used such sources to a lesser extent in the second crisis, but in both cases the reliance was not great. [See Graph 31]

Oil industry sources comprised 17% of all identifiable sources. Reliance on the oil industry for information increased from 16% in the first crisis to 18% in the second. CBS turned to oil industry sources the most in the first crisis, NBC the least. While the industry increased proportionately as a source from the first to the second crisis, it declined as a spokesman. [See Graphs 22 and 32]

A similar divergence may be seen with regard to sources from OPEC. OPEC sources were much more prominent as sources than as spokesmen. While only 2% of all spokesmen were from OPEC, 6% of all identifiable sources were OPEC sources. As a source OPEC increased slightly in importance from the first to the second crisis though it declined as a spokesman. In the first crisis ABC used OPEC sources the least, NBC the most. In the second crisis, CBS used OPEC sources the least, ABC the most. [See Graphs 23 and 33]

The man-in-the-street was a source about 5% of the time. As with the data on the man-in-the-street as spokesman, the most interesting change in the coverage of man-in-the-street was on ABC. In the first crisis, ABC used the man-in-the-street as a source more than either of the other two networks. In the second crisis, ABC was at the other extreme, using such sources the least of the three networks. [See Graph 34]

The remaining group of identifiable sources came from the several disparate groups that have been aggregated under the heading "other" sources. Such sources accounted for 16% of all identifiable sources. They were of much greater importance in the first crisis, where they accounted for 20% of all identifiable sources, than in the second, where they accounted for only 10%. [See Graph 35]

Graph 30

Percentage of Identifiable Sources From Congress
*[100% = all identifiable sources
of subjects on a network
in one crisis]*

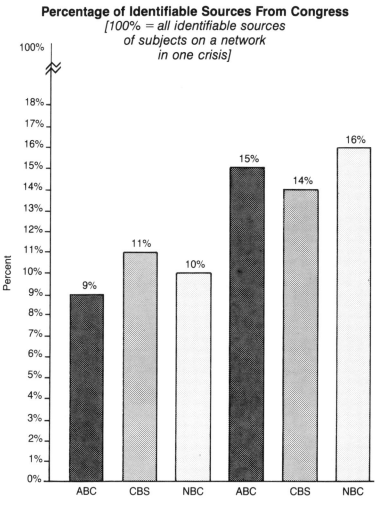

Graph 31

Percentage of Identifiable Sources From State and Local Government

*[100% = all the identifiable sources
of subjects on a network
in one crisis]*

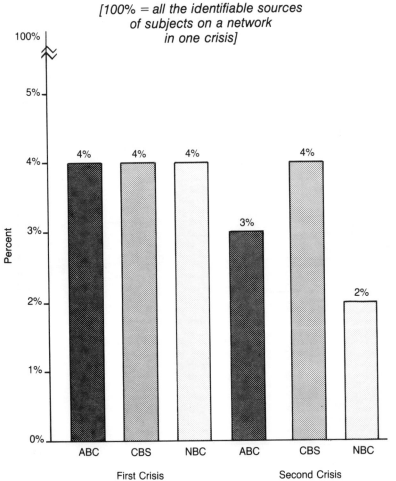

Graph 32

Percentage of Identifiable Sources From the Oil Industry
*[100% = all identifiable sources
of subjects on a network
in one crisis]*

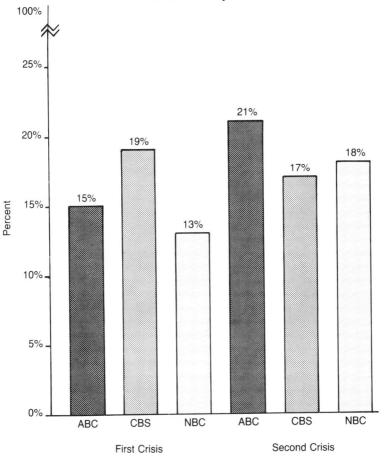

Graph 33

Percentage of Identifiable Sources From OPEC
*[100% = all identifiable sources
of subjects on a network
in one crisis]*

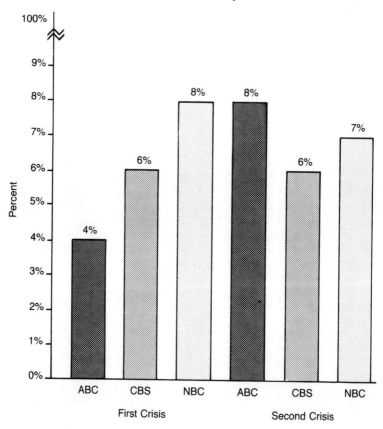

Graph 34

Percentage of Identifiable Sources Presented As "Man-in-the-Street"
[100% = all identifiable sources of subjects on a network in one crisis]

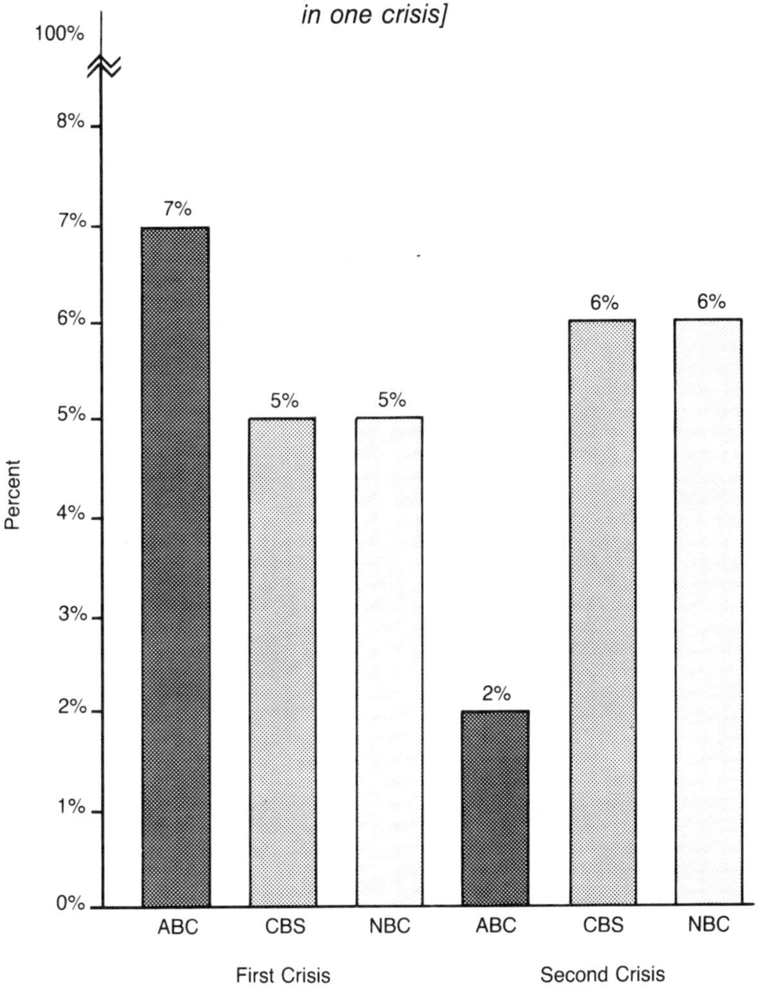

Graph 35

Percentage of Identifiable Sources in Remaining Categories—Here Aggregated as Other*
*[100% = all identifiable sources
of subjects on a network
in one crisis]*

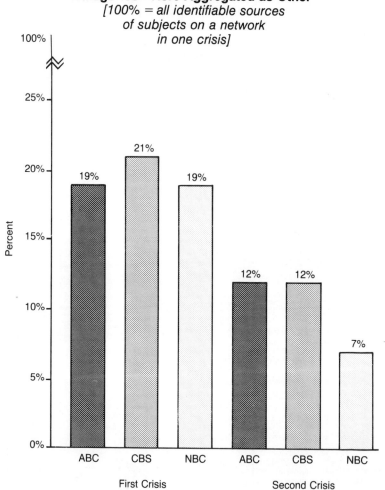

*Includes sources of non-oil industries and businesses, labor, consumer and environmental sources, experts in economics, etc, and those sources not fitting in any category provided the researcher.

Chapter IV

For the enterprising journalist, the subject of the oil crises posed infinite story possibilities. The enormous scope of the issues involved, the universal significance of the crises and the ubiquity of the resultant problems permitted virtually any conceivable tack in approaching the subject. Thus, it is interesting to examine the networks' coverage in view of this tremendous latitude. Should the focus be the international angle or the hometown ramifications, the economic or the political? Certainly the government's role in the crises could not be ignored. But was the government likely to deliver the country from the crises, or was it actually the culprit?

In fact, all of these points of view found their way into the coverage. The relative emphasis varied from one network to the next, however, and those differences will be briefly examined in this chapter. Because Volume I, published separately, addresses questions regarding the content of the oil crisis coverage, these issues will not be explored in depth here.

Around 83% of all oil crisis stories focused primarily on domestic events. The remaining 17% focused primarily on foreign or international occurrences, in particular those involving petroleum production and pricing in OPEC countries. As noted in the methodology [See Appendix], stories about the energy crises of other countries, the Arab-Israeli war and the Iranian revolution (except where such events were presented as directly affecting oil production) were excluded.

The first crisis was covered with a higher percentage (19%) of foreign stories. NBC had the highest percentage of stories which focused on foreign events, with 22%, and ABC the lowest, with 16%. Foreign stories received less attention in the second crisis—around 13%. ABC again put the least emphasis on foreign events [See Graph 36]

Distinguishing between economics and politics can be difficult because the two areas overlap. Nonetheless, such distinctions can be made, allowing for a fair number of stories that combine both points of view, and those that fit neither category (coded "other"). When the focus of stories was analyzed [See Appendix for coding rules], the first crisis proved to have been equally divided in focus between politics and economics, while the second crisis was 50% more economic in focus than political. CBS had by far the greatest emphasis on the economic angle, with 48% more stories primarily economic in focus than political. ABC had 9% more economic than political, while NBC had 11% more political than economic stories. Interestingly, the percentage of stories that combined an economic and political focus increased substantially in the second crisis. [See Graph 37]

As the measures of spokesmen and sources showed in Chapter III, the government (federal, state and local) played a major role in oil crisis coverage as a source of information. Not surprisingly, the government was also the subject of extensive coverage. In 60% of all stories, reference was made to government policies or to the government's involvement in the oil crises. In the first crisis, ABC mentioned government policies most often, in over three-quarters of its stories. By contrast, CBS mentioned government policies in just over half of its first crisis stories. In the second crisis, ABC and CBS nearly agreed in the frequency with which they referred to the government's involvement in the oil crises, while NBC referred to the government considerably more often. [See Graph 38]

When government policies were mentioned, they were generally placed in either a favorable or unfavorable light. There was yet a third category to accommodate those stories

Graph 36

Relative Focus on Foreign and Domestic Events/Affairs

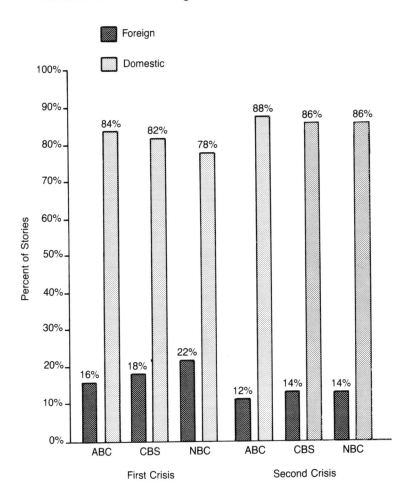

■ Foreign
▢ Domestic

Percent of Stories

First Crisis: ABC 16% / 84%, CBS 18% / 82%, NBC 22% / 78%

Second Crisis: ABC 12% / 88%, CBS 14% / 86%, NBC 14% / 86%

Graph 37

Relative Focus on Political and Economic Issues in Oil Crises Coverage

*[100% = total number of stories
broadcast by a network
during one crisis]*

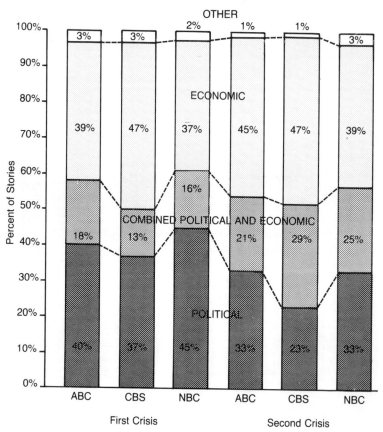

Graph 38

Reference to Government Policies or Involvement in Oil Crises

[100% = Total number of stories broadcast by network during one crisis]

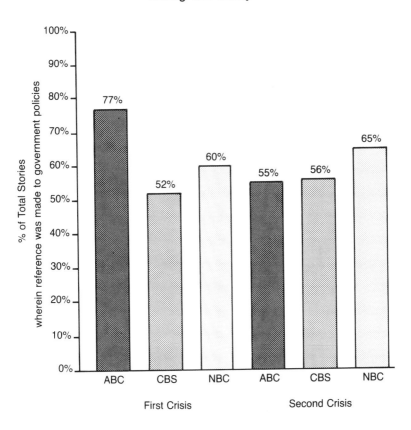

in which the government was criticized for not enough involvement in the oil crises—for example, stories about calls for the government to bring down the price of gasoline. Such stories were unfavorable in the sense they were critical of the government's policies, yet were favorable in the sense they advocated more rather than less government involvement. Thus, they were tallied separately, as "not enough government involvement". Finally, some stories mentioning the government portrayed its policies in a way that would not fit into any of these three categories, or alternately, combined several at once. These stories were classified as "other."

ABC was the most favorable in its portrayal of government policies, and the least unfavorable. CBS was the least favorable in its treatment of government policies, though at the same time the most likely to criticize the government for not enough involvement. [See Graph 39]

The first crisis contained a higher percentage of stories with a favorable portrayal of the government—66% as opposed to 61%. NBC and ABC were tied for the highest percentage of favorable stories in the first crisis (67%), while CBS had the least (62%). In the second crisis, NBC had the fewest favorable stories (58%), while ABC had a very high favorable percentage—74%. [See Graph 40]

Graph 39

Network Portrayal of Government Involvement or Policies Toward Oil Crises

*[100% = all the stories referring
to government policies
on a network]*

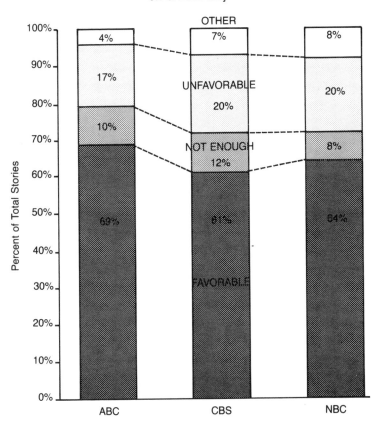

Graph 40

Network Portrayal of Government Involvement or Policies Toward Oil Crises

*[100% = all stories referring to government
policies on a network
during one crisis]*

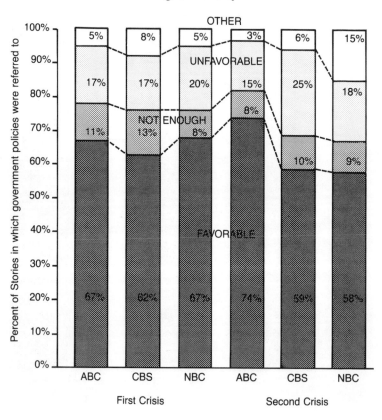

Appendix

Methodology

This study would have been impossible were it not for the Vanderbilt Television News Archive. Since 1968, Vanderbilt has been videotaping the evening news broadcasts of ABC, CBS and NBC. In addition, it publishes monthly abstracts and an index through which researchers can identify and place orders for stories of interest. (Since 1979 The Media Institute has used the Vanderbilt Archive as the source for data on business and economic stories, which the Institute publishes in its fortnightly monitor, *The Television Business-Economic News Index*.) Vanderbilt then compiles videotapes of those stories and loans out the tapes for a modest fee.

Thanks to the excellent services the Vanderbilt Television News Archive provides, The Media Institute was able to identify and obtain videotapes of every network evening news broadcast relevant to this study, a total of 39½ hours of germaine tape.

Criteria for selection of stories

The Vanderbilt Index was used to identify oil crisis stories. All stories appearing under "Energy Crisis" were included, with the following exclusions:

1. *Alternate Fuels.* Only stories pertaining to crude oil and refined products, e.g. gasoline, heating oil, aviation fuel,

and kerosene. Thus, stories concerning oil shales, synfuels, natural gas, coal, solar, geothermal, and nuclear energy were excluded.

2. *Foreign events only indirectly related to the oil crises in the U.S.* Thus, stories about the Arab-Israeli War, the Iranian Revolution, and the energy crises in foreign countries were excluded, unless they referred to the production of petroleum for export to the U.S. If a story mentioned oil production in the context of a longer story about other matters, as in a story about the Iranian Revolution that makes mention of cutbacks in production, then the story was considered to begin and end with the direct references to oil.

3. *Bureaucratic shuffles.* Stories noting, for example, that the Federal Energy Administration had had a change of command were excluded, unless some substantive policy matters were discussed in the story.

4. Stories about collisions, accidents, safety, or pollution were excluded.

5. *Editorials,* such as Howard K. Smith's "Commentaries" on ABC, Eric Sevareid's "Analysis" on CBS, and David Brinkley's "Journal" on NBC, were excluded.

Researchers determined that the coverage of oil crisis stories began in earnest in October 1973 (the OPEC embargo was declared October 17), and tapered off in May 1974. The oil crisis again became a sizeable story on the evening news in November 1978, and did not wane until August 1979.

The Coding Process

The 39½ hours of videotape were analyzed by The Media Institute researchers employing a quantitative and objective technique known as content analysis. This is a systematic and reliable method for organizing communication content into various categories. The coding sheet contains preestablished categories into which all information is coded.

The process of coding is analogous to a survey technique in many respects. The coder must answer a variety of questions by viewing the film several times, timing certain segments with a stopwatch, and applying the decision rules to the particulars of each story. Checks on the reliability of the coders were conducted, and exceeded acceptable minimums.

A separate code sheet is devoted to each story. Because several related reports often follow consecutively in a newscast, it is sometimes difficult to determine when one story has ended and the next begun. The coder therefore referred to the Vanderbilt Abstracts, wherein the beginning of each new story is indicated by a notation of the time at which it was originally broadcast.

A number of questions are designed to establish the identity of the story before it is coded for content. Vanderbilt greatly facilitates this effort by electronically coding its videotapes so that the network, date, and time of broadcast (to the nearest ten seconds) is displayed across the top of the screen.

Within any one story, matters not directly related to the oil crises were at times discussed. To avoid muddying the waters, researchers deleted from the total time given each story any tangential material that continued for thirty seconds or longer without mention of matters germane to the study.

After the coder noted the network, the date of broadcast, and the times the broadcast had begun and ended, he noted what production technique and what sort of graphics, if any, were used. These two questions required the coder to select the most sophisticated choice appropriate. Thus, a story that begins with an anchorman followed by a reporter standing in front of the White House is considered a "stand-upper"; if the story then includes videotape, it is considered a "videotape" story. Likewise, if both graphs and number overlays are used, graphs take precedence, since they generally represent a more elaborate presentation.

The remaining questions required the coder to select answers that best described the story. In those few instances where the information did not fit into the general categories, an "other" category was provided.

The foreign/domestic focus question required the coder

to judge whether a story was primarily concerned with international or domestic events. Note that most stories with a foreign focus were excluded by the criteria for selecting stories (see above).

The political/economic focus question determines whether a story was presented as a policy struggle between various political contenders, or as an economic issue affecting supplies, prices, production, consumption or distribution. The political/economic focus of a story was determined in accordance with the following rules.

Political or Economic?

For the purpose of this study, "political" refers to the pluralistic *process* of government. Thus, stories that are predominantly concerned with government process or the effects on government process are defined as "political." So defined, the election process, analysis of political popularity, poll-watching, legislative propositions and debate, political party maneuvering, administrative process, bureaucratic manipulation, foreign policy and diplomacy are "political" stories.

For the purpose of this study, "economic" refers to the activity of the "market." Stories predominantly concerned with, or emphasizing some effect on supply, demand, production, distribution, consumption, prices, surpluses, or shortages are considered "economic."

If a story is presented as equally economic and political then the category "combined political/economic" applies.

If a story is presented from neither an economic nor a political viewpoint, then the category "other" should be marked and reservations written on the back of the code sheet.

Obviously, many issues can be approached from either a political or an economic viewpoint. For example, "windfall profits," when discussed from an angle that probes the possibility of legislative passage, is a political story; when the same topic is discussed from an angle that examines the legislation's effect on supplies or prices, the topic becomes "economic."

Favorable or Unfavorable?

The portrayal of the government's role was measured with a choice of options. As with the previous two questions, coders were required to select the best answer. For this study, answers were aggregated as follows:

1. Favorable—the combined stories coded as "favorable due to domestic policy" and "favorable due to foreign policy."

2. Unfavorable—the combined stories coded as "harmful: allocation system is bad," "harmful: government discourages domestic production," "harmful due to domestic policy" and "harmful due to foreign policy."

3. Not enough government involvement.

4. Other—the combined stories coded either as "more than one of the above" or "other."

5. Government's role not measured.

A preliminary survey of the videotapes, employing a modified form of emergent analysis, had resulted in the identification of all principal segments of society to appear in news coverage of the crises. For those segments of society, such as the government, which appeared or were quoted frequently in the news, further subcategories, such as Congress or state and local government, were established. This same process also resulted in the identification of the major categories and subcategories of coverage. As a result, the researchers had a list of 60 categories that accommodated all sources of information, and a second list of 200 categories that accommodated all subjects discussed.

In viewing each story, coders noted each subject that was raised, and the source of such subjects. An individual story might be adequately summarized by one such source-subject combination, or might require two dozen or more such combinations, depending on how complicated a story was and how much information it contained.

In those instances where the source was on camera while he provided information, the length of time he spoke was timed with a stopwatch. The total time a spokesman was seen and heard speaking in a story was measured, excluding interruptions, such as queries from interviewers. In such cases, a source was also counted as a spokesman.

The data collected in this fashion was then analyzed with the assistance of an IBM 3033 at the George Washington University Center for Computing in Washington, D.C. In this way the data was aggregated and disaggregated in a wide variety of ways.